Heinemann Experimental Chemistry Series

General Editors:

A. J. Mee
Martin Rogers, *Headmaster, Malvern College*

Advisory Panel:

Professor E. H. Coulson, *Organizer, Nuffield Advanced Chemistry Project*
Professor H. F. Halliwell

10
Radiochemistry

Students' Text

Heinemann Experimental Chemistry Series

1—Structure, Equilibrium, and Kinetics
A. J. Mee and Derek Bradford

***2—Electrochemistry**
D. Mansfield

***3—Chemical Analysis, Chromatography, and Ion Exchange**
R. T. Allsop and J. A. D. Healey

4—Transition Element Chemistry
J. N. King and T. McManus

5—Atomic and Molecular Masses
Derek Bradford

6—Colloids
D. Mansfield

7—Carbon Chemistry
A. G. Rayden

8—The Periodic Table
Claude Evans

***9—Metals and Alloys**
F. W. J. Pargeter and M. R. Mander

***10—Radiochemistry**
D. Mansfield

Note: only those titles marked * have been published at the time of publication of the present volume.

Heinemann Experimental Chemistry Series

10
Radiochemistry

Students' Text

D. Mansfield, B.Sc.

Head of the Chemistry Department
Harvey Grammar School, Folkestone, Kent

Heinemann Educational Books

Heinemann Educational Books Ltd

London Edinburgh Melbourne Auckland Toronto
Hong Kong Singapore Kuala Lumpur
Ibadan Nairobi Johannesburg
Lusaka New Delhi

ISBN 0 435 65968 5

First published 1974

Published by Heinemann Educational Books Ltd
48 Charles Street, London W1X 8AH

Printed in Great Britain by
William Clowes & Sons Limited
London, Beccles and Colchester

Preface to the Series

The richness of the British tradition of science teaching lies in its experimental work. Nowhere in the world has there been for so long such an emphasis on experiments performed both by teachers and students. The books in this series have been written to sustain that tradition and to inject new life into it.

Much of the practical work in chemistry which was done by students in the past consisted merely of the repetition of experiments, the results of which were already known, or in the preparation of compounds by following a rigid set of instructions. In either case there was little room for thought on the part of the student. Modern chemistry teaching seeks to improve on this situation by emphasizing that the aim of practical work is to discover something, not merely to copy what has been done before. Of course, experimental work at this level is not, in fact, research, but there is no reason why it should not be presented as if it were. To the student it *is* something new that he is finding out.

These books have been designed to cover the whole area of advanced school chemistry and introductory courses at colleges and universities. A series of books, each centred on a specified topic, offers considerable advantages over a single volume. It is possible to include more material and to present it in a very flexible way. Teachers can choose those volumes which suit their own particular course and students are not stuck with a heavy volume containing a good deal of irrelevant material. Moreover, each book can be written by a teacher, expert in a particular field.

What is new in this series is the method of treatment. The approach is based on discovery by the students. They are given sufficient information to perform the experiments in an investigational manner and to think about them, but not so much that the performance becomes a meaningless ritual. The experimental work is not an end in itself. The experiments serve some definite purpose in relation to the theory work. By means of questions students are led to examine the results of their work, not only with regard to the degree of accuracy they may have attained but to deduce theoretical points arising from the experiments. In this way the practical work becomes an integral part of the chemistry course and not a mere adjunct to it.

We have attempted to include a number of modern experiments which are new to the school laboratory. Teachers will recognize the many 'classical' experiments which also appear, although these have been treated with a different emphasis.

The books are intended to allow students to proceed at their own pace. This does not mean, of course, that the teacher is absolved from the responsibility of over-seeing the student's work, but it does mean that students can get on without having a lecture period beforehand in which the teacher explains what has to be done in the practical session. It is hoped that the use of the books will foster independence of thought among students and will thus enable the teacher to assume his proper role of counsellor and guide rather than instructor.

For each *Students' Text* a *Teachers' Guide* has been prepared. It gives hints about the conduct of the experiments and the conclusions that can be drawn from them, and contains the answers to most of the questions posed in the *Students' Texts*. The list of materials required for each experiment is provided in the *Students' Texts*.

We cannot hope to have produced a series of books which is completely free from error and we hope that users will inform us of any lapses. It would be foolish to claim, too, that the experiments described are incapable of improvement and we should be glad to hear of any way in which this can be done.

Specific acknowledgment of the considerable amount of help received in the writing of these books will be found in the Introduction to the individual volumes. We are, however, deeply indebted to Prof. E. H. Coulson and Prof. H. F. Halliwell for acting as consultants in the preparation of the series as a whole. They gave very valuable advice from which all the books have benefited.

The study of chemistry through experiments is not only great fun, it is an excellent education. We hope that the teaching of experimental chemistry will be encouraged and enriched by this series.

A. J. Mee
Martin Rogers

1974

Contents

Preface to the Series v

Introduction 1

1: The Detection of Radiation from Radioactive Materials

The Geiger-Müller Tube 3

1.1 Becquerel's Experiment 4
1.2 Autoradiography 5

2: The Stability of the Nucleus

2.1 Effect of Concentration on Activity 8
2.2 Preparation of a Radioactive Source and Investigation of the Random Nature of Radioactive Decay 9
2.3 An Analogue of Radioactive Decay using Dice 10
2.4 Ease of Absorption of Radiation 14
2.5 What is the Law relating Intensity of Activity and Distance from a Source? 15

3: Some Separation Techniques

3.1 Investigation of the Growth and Decay of Protactinium-234 16

The Radioactive Decay Series 18

3.2 Separating Radon-220 and Finding its Half-life 19
3.3 Separating Lead-212 and Finding its Half-life 22
3.4 Separating Bismuth-212 and Finding its Half-life 23
3.5 Separating Thallium-208 and Finding its Half-life 24

4: Some Tracer Techniques

4.1 Is there an Equilibrium between Solid Lead Chloride and its Saturated Aqueous Solution? 26
4.2 Investigation of the Reaction between Solutions of Thorium(IV) Nitrate and Sodium Oxalate 27

4.3 Investigation of the Distribution of Iodine between Tetrachloromethane and Water 28
4.4 Finding the Equilibrium Constant of a Redox Reaction 29
4.5 Investigation of the Uptake of Radioactive Materials by Plants 31
4.6 Investigation of the Separation of Sugars using Paper Chromatography 32
4.7 Investigation of the Photosynthetic Activity of a Plant using Carbon-14 as a Tracer 33

Appendix 36

Logarithms 38

e^{-x} from $x = 0.00$ to $x = 5.50$ (in movements of 0.01) 40

Introduction

Radiochemistry is a young and exciting branch of science that is daily making greater contributions to our understanding of chemical processes. In working through this book you will be carrying out a series of experiments from which you will learn much about the nature of radioactive materials, the techniques used in their investigation, and the ways in which scientists utilize their properties to obtain solutions to a wide range of problems.

So that this work may be carried out in a safe and efficient manner the following **precautions** must be observed.

(1) All closed sources must be handled with an approved tool and, if directional, they must not be pointed at the experimenter or any other person.

(2) When using open sources a laboratory coat (kept for this purpose and laundered separately from other clothes) **must** be worn in conjunction with a pair of disposable polythene gloves. All such work must be carried out over a tray lined with material as directed by the teacher. Every precaution against contamination by spillage must be observed. All solid waste must be placed in the special waste bin assigned for this purpose and all liquid waste must be disposed of in the bottle specially provided.

(3) Forceps must be used at all times when handling radioactive materials, such as planchettes. The contamination of instrument controls, Geiger–Müller (G–M) tubes etc., must be avoided by handling these with a paper tissue when working with open sources.

(4) Only glassware that is reserved for radiochemical work and has been appropriately labelled should be used.

(5) Radioactive solutions must never be pipetted by mouth; a plastic syringe must be fitted to the pipette with rubber tubing to provide suction (see Figure 2.3).

(6) No departure from the instructions given for the performance of the experiments may be made without the direct approval of your teacher.

If your laboratory work in radiochemistry is not supplemented by theory lessons, you should read a suitable book concurrently with your experimental work. Such a book is *The Restless Atom* by Alfred Romer (Heinemann Educational Books; Science Study Series, No. 10).

To understand the experimental work this book contains, you should be familiar with the following landmarks in the development of radiochemistry.

1896: Henri Becquerel discovered atomic radiations by their ability to cause fogging of photographic plates excluded from visible light. Marie and Pierre Curie shortly afterwards isolated two new powerfully radioactive elements, polonium and radium, from the mineral pitchblende. It was later shown that if a beam of radiation from radium was passed between a pair of electrically charged plates (or between the poles of a magnet) it split into three components—positively charged α-particles, negative β-particles and uncharged γ-electromagnetic radiation.

1897: J. J. Thomson discovered the electron as a constituent of atoms from experiments with gas discharge tubes. He postulated the 'plum-pudding' model of the atom according to which atoms were jelly-like substances in which electrons were embedded.

1905: Albert Einstein published his 'Special Theory of Relativity' in which it was shown that matter and energy are interconvertible according to the equation $E = mc^2$.

1909: Hans Geiger and Ernest Marsden showed that when thin gold foil is bombarded with α-particles, a proportion of these are scattered through unexpectedly large angles.

1911: Ernest Rutherford used the results obtained by Geiger and Marsden to develop his model of the nuclear atom.

1912: J. J. Thomson discovered that neon consists of more than one isotope using a mass spectrograph.

1920: Ernest Rutherford predicted the existence of the neutron.

1921: Ernest Rutherford discovered that nitrogen can be transmuted into oxygen by bombardment with α-particles.

1930: P. A. M. Dirac predicted the existence of the positron.

1932: J. Chadwick demonstrated the existence of neutrons experimentally by showing that they would dislodge α-particles from the nuclei of certain atoms, such as boron, by bombardment. Carl Anderson discovered the existence of positrons in cosmic radiation. Irene Curie and F. Joliot discovered that many light elements could be transmuted by α-particle bombardment and that positrons were often formed in the process.

1: The Detection of Radiation from Radioactive Materials

The earliest methods of detecting radioactivity, as we shall see later, involved the fogging of photographic plates. Although this can be developed to give semi-quantitative measurements (e.g. the lapel badges worn by those working in establishments which use potentially dangerous levels of radioactivity), this is not sufficiently accurate for our purposes.

Radioactive emissions will ionize the air in gold-leaf electroscopes and discharge them. The rate of discharge was used by early experimenters (to measure radioactivity), but the cumbersome and unreliable nature of the apparatus has caused this technique to fall largely into disuse after the invention of the Geiger–Müller tube (G–M tube).

The Geiger-Müller Tube

The construction of a G–M tube is shown in Figure 2.2. A liquid counter is merely a special form of G–M tube constructed to count the radioactivity of liquids in which the G–M tube proper is surrounded by a second concentric tube to hold the liquid under test. The G–M tube itself contains a gas at low pressure (often a halogen or organic compound) and two electrodes which are maintained at a suitable potential difference. (Note : the manufacturer's literature should be consulted to determine the correct working voltage for any G–M tube you use.) When any radiation passes into this chamber it causes the gas to ionize and allow a current to pass between the electrodes. If the G–M tube is connected to an instrument called a scaler, this counts the number of times such ionization results in a current passing and this gives a quantitative measure of level of activity from the source. If the G–M tube is connected to an instrument called a ratemeter, this information concerning the number of ionizations taking place is electronically differentiated and presented as a rate with respect to time. The method of operation of the different models of such apparatus available varies considerably and the manufacturer's booklet should be consulted if there is any doubt about usage. (Note : it is essential to handle the controls of these instruments through paper tissues when using open sources to avoid the possibility of contamination.) So that background radiation is absorbed and not counted in radiochemical experiments, a cylinder of lead can be placed around the G–M tube and holder. Such an arrangement is called a 'lead castle'. It is not essential to use a lead castle for the experiments in this book.

1.1 Becquerel's Experiment

The experiment which you are about to carry out is the classical one by which Becquerel discovered radioactivity in 1896. He found that uranium compounds emitted a penetrating radiation similar in properties to X-rays. Two years later Mme Curie found that thorium(IV) compounds behaved in the same way.

Apparatus and Materials

Sealed polythene bag containing solid thorium(IV) nitrate or uranyl nitrate
Envelope-wrapped X-ray film
Access to dark room illuminated with safelight fitted with 6B brown Wratten filter and containing:
- dish of X-ray developer
- dish of X-ray fixer
- washing tank (with running water)

photographer's plastic tongs
line of clips (preferably of plastic) on which to hang developed film to dry
clock with seconds hand
thermometer
key, coin, or other suitable metallic object

Procedure

Set up the apparatus as shown in Figure 1.1 and leave undisturbed for about forty-eight hours.

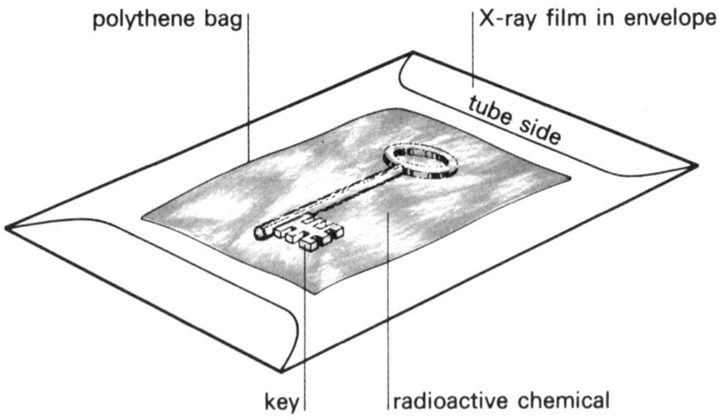

Figure 1.1 The 'Becquerel experiment'

Open the envelope in the dark room and remove the film, using the plastic tongs. Note the temperature of the developer and use the manufacturer's instructions to determine from this the correct development time. Place the film in the developing tank and use the tongs to agitate the solution at frequent intervals, taking care to avoid scratching the film. When development is complete, transfer the film to the washing tank for at least five minutes. Place the film in the fixing solution for the time specified by the manufacturer. When this is over, remove the film and place it in the washing tank for a further five minutes before hanging it to dry. If a hair dryer is available, this can be used to accelerate the drying process.

1.2 Autoradiography

As the title of this experiment implies, we are going to use materials which are themselves radioactive to produce their own images on a photographic film.

Apparatus and Materials

Photographic requirements as specified for Experiment 1.1

Luminous watch dial or other luminous instrument dial

Radioactive mineral specimens such as autunite, davidite, monazite, phosphouranylite, pitchblende, terbernite

Procedure

Place each of the specimens on an envelope containing X-ray film and leave undisturbed for forty-eight hours. Develop the film using the procedure described in Experiment 1.1. If any of the films are under- or over-exposed the experiment may be repeated with the exposure time changed accordingly.

2: The Stability of the Nucleus

Your theoretical work on this topic should have made you aware that:

(a) some isotopes are stable whilst others undergo spontaneous radioactive decay;

(b) there are several modes of radioactive decay including:
α-emission of two protons plus two neutrons as a packet from the nucleus,
β^--emission of an electron from the nucleus,
β^+-emission of a positron from the nucleus, and
γ-emission of electromagnetic radiation from the nucleus.

α-emission is largely confined to the heavier atomic nuclei. To find out more about the conditions under which β^-- and β^+-emission occur, complete Table 2.1.

Table 2.1 Some isotopes of selected elements.

Atomic number	*Element*	*Mass number*	*Number of neutrons*	*Number of neutrons / Number of protons*	*Mode of decay*
1	H	1			Stable
		2			Stable
		3			β^-
4	Be	7			β^+
		9			Stable
		10			β^-
		12			β^-
6	C	10			β^+
		11			β^+
		12			Stable
		13			Stable
		14			β^-
		15			β^-
9	F	17			β^+
		18			β^+
		19			Stable
		20			β^-
		21			β^-
		22			β^-

Table 2.1 *(continued)*

Atomic number	*Element*	*Mass number*	*Number of neutrons*	*Number of neutrons / Number of protons*	*Mode of decay*
12	Mg	21			β^+
		22			β^+
		23			β^+
		24			Stable
		25			Stable
		26			Stable
		27			β^-
		28			β^-
15	P	28			β^+
		29			β^+
		30			β^+
		31			Stable
		32			β^-
		33			β^-
		34			β^-
22	Ti	43			β^+
		44			β^+
		45			β^+
		46			Stable
		47			Stable
		48			Stable
		49			Stable
		50			Stable
		51			β^+
		52			β^+

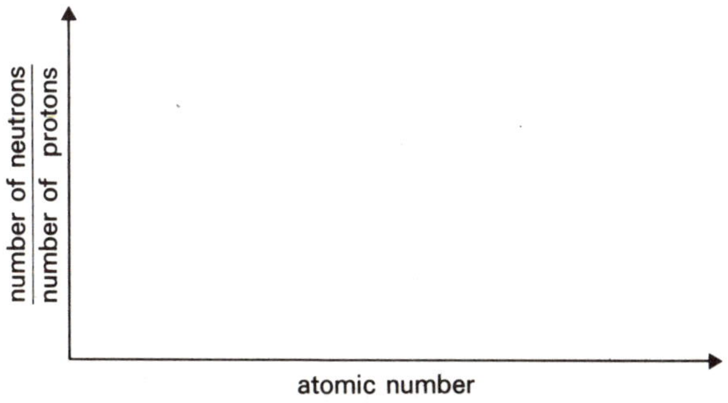

Figure 2.1

Now plot a graph of $\frac{\text{number of neutrons}}{\text{number of protons}}$ against atomic number, as shown in Figure 2.1. It will make any pattern that emerges clearer if you use different colours for stable isotopes, β^--emitters, and β^+-emitters.

Is it possible to make any generalization about the stability pattern shown?

2.1 Effect of Concentration on Activity

One would expect to find that the activity of a solution of a radioactive substance would increase with the concentration. The following experiment will enable you to find out how far this is true, and will introduce you to the technique of working with some of the special apparatus employed in radiochemical work.

Apparatus and Materials

Scaler with liquid counter and holder
Lead castle (optional)
Stop-clock
Several 100 cm³ beakers
Graduated pipettes of various sizes for making dilutions—fitted with plastic syringes as shown in Figure 2.3
1.0 M solution of uranyl nitrate

Procedure

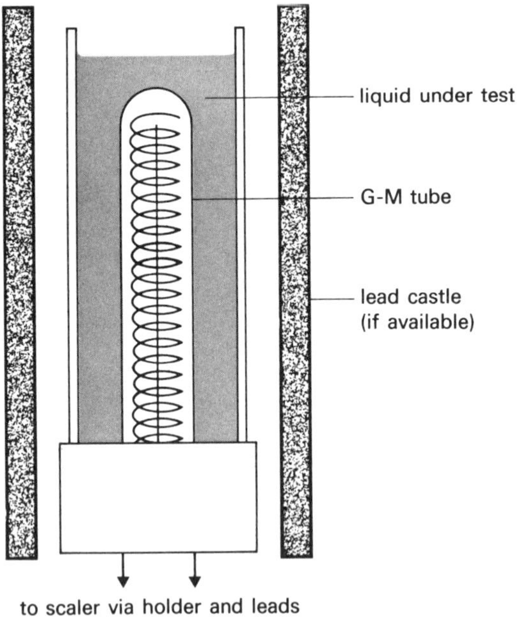

Figure 2.2 Liquid counter

Place distilled water in the liquid counter (Figure 2.2) and take the count over five minutes. Rinse the liquid counter with 1.0 M uranyl nitrate and discard the rinsings. Then fill the counter with the solution. Take the count over one minute.

Make a series of dilutions of the 1.0 M uranyl nitrate down to 0.001 M and find the activity of each.

Remember that on no account may radioactive solutions be sucked up in a pipette by mouth. The pipette must be filled with a plastic syringe as shown in Figure 2.3.

Figure 2.3 Using a plastic syringe to fill a pipette

Rinse out the liquid counter between each determination with two portions of distilled water and then with the liquid under test.

Results

Enter your results in a table as shown below.

Concentration of uranyl nitrate in mol dm^{-3}					
Count rate per minute					

Plot a graph of your results. Does the graph pass through the origin? If not, why?

2.2 Preparation of a Radioactive Source and Investigation of the Random Nature of Radioactive Decay

Radioactive sources are supplied in a number of forms, but in this experiment you will make a source for yourself from uranium oxide in a sealed planchette. This type of source is known as a 'sealed' source and once prepared is intended to be kept permanently without being tampered with. You will use the source you prepare to investigate its rate of decay.

Apparatus and Materials

Wax candle
Metal can (about 250 cm^3 capacity)
Powdered uranium oxide
Spatula (nickel)
Planchette and lid (about 1.5 cm in diameter)

Adhesive (such as Durafix)
Metal forceps to handle source
Stop-clock
Perspex square, with hole to hold source
Scaler with G–M tube and holder

Procedure

(1) *Preparation of the source*

Melt the candle carefully in the tin can over a low Bunsen flame and remove the wick. Add about half the same bulk of powdered uranium oxide and stir to form a thick paste. There will be sufficient material here to fill many sources and you may be provided with a paste prepared on a previous occasion. In this case the paste requires only careful remelting. Taking care to avoid spillage, transfer some of the paste to a planchette, as shown in Figure 2.4(a). When the paste has set, place adhesive round the rim of the planchette, and put the lid firmly in place. Leave the completed source (which should look as shown in Figure 2.4(b)) long enough for the adhesive to set.

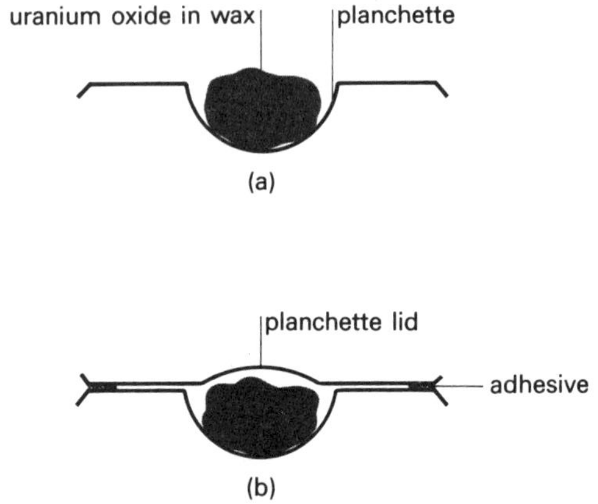

Figure 2.4 Making a β-source

(2) *Random nature of radioactive decay*

Place the source in a perspex holder and clamp the G–M tube some distance above it, as shown in Figure 2.5. Adjust the distance between the G–M tube and the source until a count rate of about 500 per minute is obtained.

Keeping this distance constant, make 100 counts of the source, each over a one minute interval. Record your results in a table as shown below.

Count rate per minute	450 or less	451–470	471–490	491–510	511–530	531–550	551 or more
Frequency of obtaining this count rate							

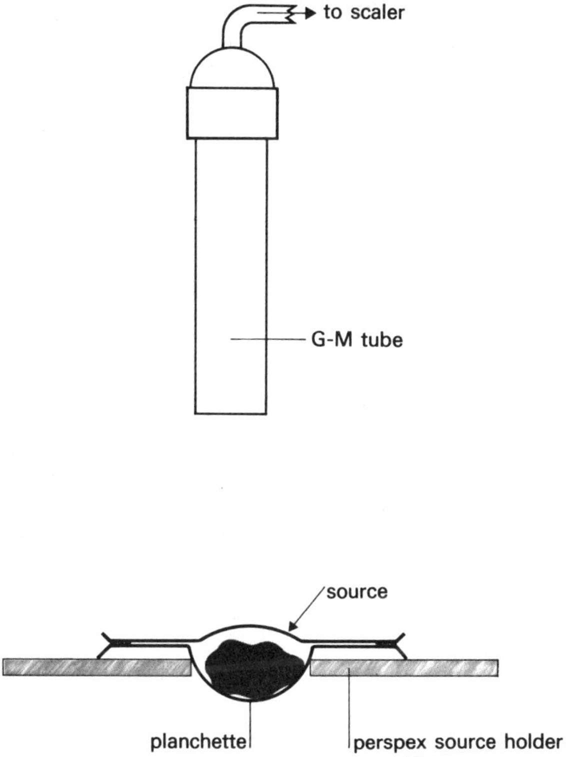

Figure 2.5 Counting the source

Plot a graph of frequency against count rate.

1. What count rate does your graph suggest is the most likely to be obtained? This is called the 'mean' count rate, and can be given the symbol m.
2. Is the variation in count rate due to any cause other than the random nature of radioactive decay?

2.3 An Analogue of Radioactive Decay using Dice

This experiment will help you to understand the random nature of radioactive decay and will enable you to deduce a mathematical expression for the 'half-life' of a random process.

Apparatus and Materials

This experiment is best done on a co-operative basis by about ten students (or pairs) each of whom will require twelve dice and a 250 cm^3 beaker. An adding machine is useful.

Procedure

Make a copy of the table shown below.

Number of dice remaining / *Number of throws*	*Game*												*Total of all games*
	A	*B*	*C*	*D*	*E*	*F*	*G*	*H*	*I*	*J*	*K*	*L*	
0	12	12	12	12	12	12	12	12	12	12	12	12	144
1													
2													
3													
4													
5													
6													
7													
8													
9													
10													

Throw the twelve dice from the beaker and put to one side all those which come down 'six'. Record in the table the number of dice that remain. Replace these dice in the beaker and throw again. Again put on one side all those that come down 'six' and record the number remaining. Put those which did not fall 'six' back into the beaker and throw again. Continue doing this

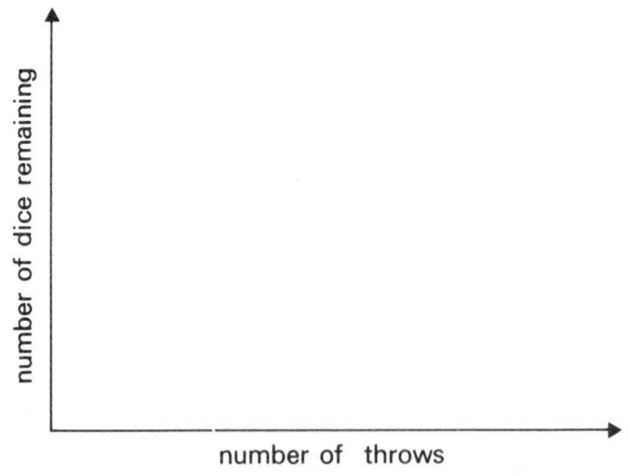

Figure 2.6

until no dice remain, or there have been ten throws. Repeat the whole experiment eleven more times and thus complete the table. Add the totals across the table, so obtaining the results that represent the throwing of 144 dice initially. Pool the results of all members of the class and obtain the grand totals for each number of throws. Plot these grand totals on a graph as in Figure 2.6.

Results and Discussion

(a) From your graph obtain the number of throws that must be made before half of any particular number of dice will be lost, recording your results as follows:

Number of dice remaining (N)	*N/2*	*Number of throws to lose N/2*
	Average	

The result gives what may be termed the 'half-life' of the dice.

(b) More dice are lost in the earlier than in the later throws. One would expect the same fraction (one-sixth) of the dice remaining to be lost at each throw. If N is the number of dice and t is the number of throws, then

rate at which dice are lost $= k \times$ number of dice remaining

where k is the proportionality factor, or

$$-\frac{\mathrm{d}N}{\mathrm{d}t} = kN$$

This can be re-arranged in a form suitable for integration as follows:

$$-\frac{\mathrm{d}N}{N} = k\,\mathrm{d}t$$

On integration this gives

$$-\ln N = kt + C$$

where C is the integration constant. To evaluate C we note that when $t = 0$ all the dice remain in play and $N = N_0$, where N_0 is the original number of dice. Thus

$$C = -\ln N_0$$

and

$$-\ln N = kt - \ln N_0$$

or

$$\boxed{\ln \frac{N_0}{N} = kt} \qquad (1)$$

For the special case of the half-life ($t_{1/2}$),

$$\frac{N_0}{N} = 2$$

and

$$\ln 2 = kt_{1/2}$$

or

$$t_{1/2} = \frac{2.3 \times 0.301}{k} = \frac{0.693}{k} \qquad (2)$$

Application of these Equations to the 'Dice Decay' Situation

If we started with six dice initially, we would expect five to remain after the first throw if the dice behaved perfectly. Substituting these figures in equation (1) we get

$$\ln \frac{6}{5} = k$$

or

$$k = \ln 1.2 = 0.182$$

Substituting this for k in equation (2) we have

$$t_{1/2} = \frac{0.693}{0.182} = 3.81$$

How well does this result agree with the value you obtained experimentally in (a)?

2.4 Ease of Absorption of Radiation

Because of the nature of the various types of radioactive radiation one might expect that they would be absorbed to different extents by materials. Predict which of the three types of radiation, α-, β-, and γ-, would be absorbed most readily and which least.

Apparatus and Materials

Scaler with thin-walled G–M tube with holder
Holder for sources
α-, β-, and γ-sources with handling tool
Lead castle (optional)
Stop-clock
Range of absorbing materials (e.g. tissue paper, thin cardboard, thick cardboard, aluminium foil, lead foil, lead sheeting)

Procedure

Place a γ-source at a suitable distance from the G–M tube to give a count rate of about 1000 per minute. Obtain a value for the count rate per minute.

Find values for the percentage activity absorbed with each absorber in turn placed between the source and the G–M tube. Repeat the experiment using an α- and then a β-source.

Discussion

What conclusion can you reach regarding the relative ease of absorption of these radiations? Does it agree with your prediction?

2.5 What is the Law relating Intensity of Activity and Distance from a Source?

If a particular value for the activity (as determined by count rate) is observed at a certain distance from a source, you would expect to obtain an identical activity at any point on the surface of a sphere which has a radius, *r*, with the source at its centre. What have you assumed in making this statement? How is the surface of a sphere related to its radius? How then would you expect the activity of a source to vary with the distance from the source?

Apparatus and Materials

Scaler with thin-walled G–M tube and holder
Holder for sources
α-, β-, and γ-sources with handling tool
Meter rule
Stop-clock

Procedure

Place a γ-source at a suitable distance from a G–M tube to give a count rate of about 5000 per minute. Obtain a value for the count at this distance and measure the distance between the source and the G–M tube. Now find the count rate at various measured distances from the source. Count rates between 5000 and 500 per minute should be used. Repeat the experiment using an α- and then a β-source.

Results

You could check if your reasoning above were correct by plotting count rate against 1/distance^2. However, you probably had some difficulty in measuring the distance from the source to the G–M tube accurately, and you will feel less confident in these values than in those of the count rate. When the distances are squared the errors are, of course, greatly magnified. You can overcome this difficulty by plotting instead count rate $^{1/2}$ against 1/distance.

Questions

1. Do your results agree with your prediction?
2. If there is any variation how can you explain it?

3: Some Separation Techniques

3.1 Investigation of the Growth and Decay of Protactinium-234

Naturally occurring uranium consists principally of a mixture of isotopes of mass numbers 235 and 238. Commercially available samples of uranium compounds have been processed to remove virtually all the already small proportion of uranium-235. Uranium-238 undergoes sequential decay before becoming a stable isotope of lead, the initial steps in the decay sequence being

$$^{238}_{92}U \xrightarrow{\alpha} {}^{234}_{90}Th \xrightarrow{\beta^-} {}^{234}_{91}Pa \xrightarrow{\beta^-} {}^{234}_{92}U$$

In the experiment below you will use two methods to separate protactinium-234 from its precursors in the series. These methods are (1) solvent extraction and (2) ion exchange. All the activity of these extracts can be regarded as being due to protactinium-234 because the half-life of its daughter product, uranium-234, is so long (2.4×10^5 years).

Protoactinium-234 is soluble in 4-methylpentan-2-one whereas thorium-234 and uranium-238 are not.

Apparatus and Materials

Scaler with liquid counter, holder, and lead

(1) *Solvent extraction method:*
- 10 cm³ measuring cylinder
- 50 cm³ separating funnel (with well-fitting stopper)
- 100 cm³ beaker
- 35 per cent aqueous solution of uranyl nitrate
- Concentrated hydrochloric acid

Stop-clock (or automatic timing control)

4-methylpentan-2-one (isobutyl methyl ketone)

(2) *Ion exchange method:*
- Ion exchange resin (Zeo-Karb 225)
- 50 cm³ burette with glass stop-cock
- Glass wool
- 15 per cent aqueous solution of uranyl nitrate
- 0.05 M sulphuric acid
- 2 M hydrochloric acid

Procedure

(1) *Solvent extraction method*

(a) Place 3 cm³ of 35 per cent uranyl nitrate solution and 7 cm³ of concentrated hydrochloric acid in the separating funnel and add 10 cm³ of 4-methylpentan-2-one. Shake the liquids together for about two minutes and allow them to separate again into two layers. Working quickly, run off the lower (aqueous) layer into a beaker and the upper layer, which contains the dissolved protactinium-234, into the liquid counter. Simultaneously start the stop-clock and scaler (or commence counting cycle if timer is available). The clock should run continuously and you should count for twenty second periods, with ten second intervals during which you read and re-set the scaler. You will almost certainly need assistance to record your results in the time available.

(b) Repeat the experiment, but this time pour the aqueous layer from which the protactinium has been removed into the liquid counter. This aqueous layer contains uranium-238 and thorium-234, which are decaying with the formation of protactinium. Count the liquid as before. The radiations from the uranium and thorium will be almost completely absorbed by the glass wall of the counter and by the solution itself. The graph of count rate against time is the growth curve for protactinium-234.

(2) *Ion exchange method*
Make a slurry of 5 g of ion exchange resin with water and leave it in a sealed jar overnight. Decant the water from the resin just before use and place the resin in a burette, the lower end of which has previously been packed with glass wool. Pour water into the burette above the resin (which must now be kept continuously moist) and run the liquid out from the tap, repeating the process until the emergent liquid is clear. Pass 30 cm^3 of the 15 per cent solution of uranyl nitrate through the resin, taking between five and seven minutes to do this. The cations will all be adsorbed. Pass 150 cm^3 of 0.05 M sulphuric acid, followed by 50 cm^3 of deionized water slowly through the resin to elute the uranyl cations. Dispose of the waste effluent in the liquid radioactive waste bottle. The resin now holds thorium-234 in equilibrium with its daughter product, protactinium-234. Elute the protactinium from the resin by passing 15 cm^3 of 2 M hydrochloric acid through it. (The elute contains a little protactinium and some other active isotopes and should be disposed of.) Then add a further 10 cm^3 of 2 M hydrochloric acid, running the latter directly into the liquid counter and counting it, as before, without delay.

Results

(1) *Solvent extraction method*
We wish to find how the count rate of protactinium-234 varies with time in order to calculate the half-life of the isotope. If, however, we plot the graph of count rate against time we meet the difficulty of having taken the count over an appreciable time interval but requiring this to be shown as a point on the graph. Fortunately there is but a small error in taking the time for the first count as 0, the second as thirty seconds, and so on. Record your results in this way and plot the decay curve as shown in Figure 3.1.

From your graph find the time it takes for a convenient count rate to drop to half its value. Repeat this using other points on the curve and average your values for the half-life of the isotope.

A mathematically more sophisticated method of finding the half-life is to plot the logarithm of the count rate against time and draw the best straight

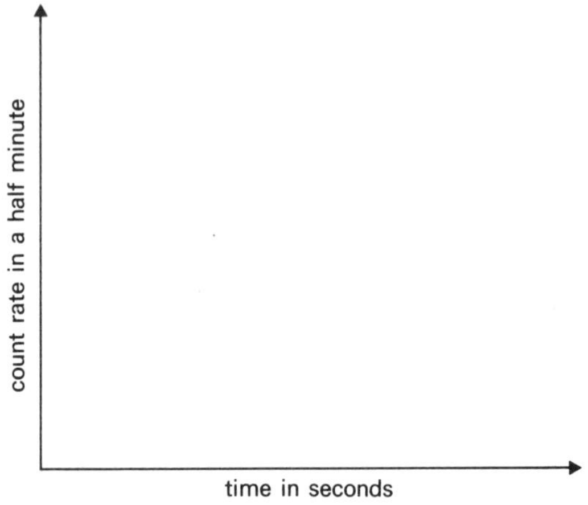

Figure 3.1

line through the points. The justification for this will be seen from Experiment 2.3.

(2) *Ion exchange method*
Plot graphs similar to those drawn in (1) and from them obtain the half-life of protactinium-234.

The Radioactive Decay Series

In addition to the family of radioactive elements formed by the sequential decay of uranium-238, two others are also found in nature. One of these starts with thorium-232, and the other with uranium-235. The mass numbers of the thorium series are all multiples of 4 and may thus be represented as $4n$. Those of the uranium-238 and the uranium-235 series can be similarly represented as $(4n + 2)$ and $(4n + 3)$, respectively. This suggests the possible existence of a fourth series of which the mass numbers can be represented by $(4n + 1)$. This series is not found in nature, but the elements in it have been artificially produced. The thorium series is shown in the scheme in Figure 3.2. All four series are shown in the Appendix (page 36). As the table in the Appendix gives the half-lives of the elements it should not be consulted until the work on this section is finished!

Experiments 3.2, 3.3, 3.4, and 3.5 are concerned with the thorium-232 ($4n$ series). In these experiments you will encounter a wide range of methods for separating individual members of such a series.

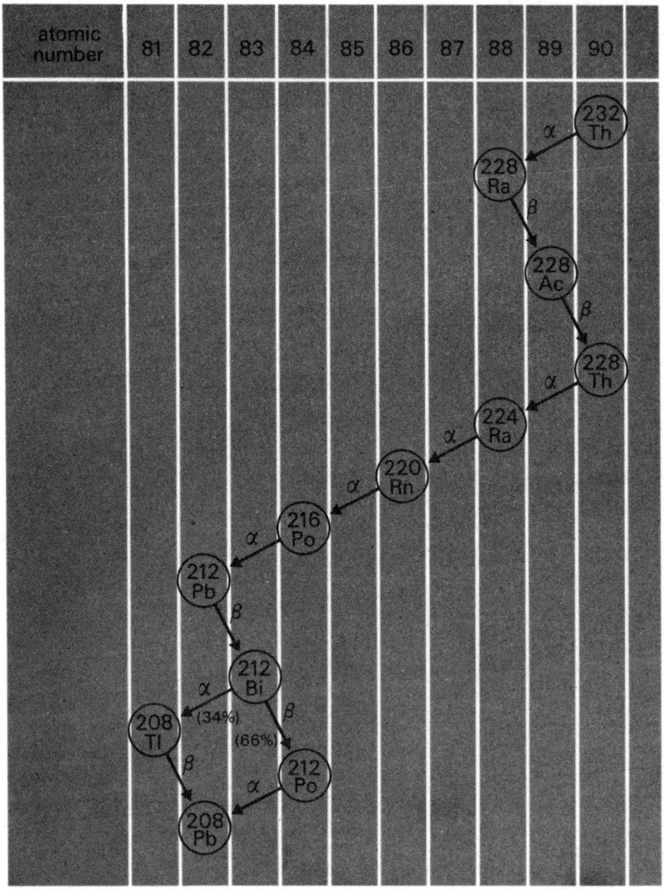

Figure 3.2 The thorium decay series

3.2 Separating Radon-220 and Finding its Half-life

Radon is the only member of any radioactive series which is gaseous under ordinary conditions. Two isotopes of radon occur in nature—radon-220 and radon-228. The half-life of radon-228 is nearly seven years, while that of radon-220 is very much shorter. Hence the activity of the gas freed from thorium compounds can be taken as being due entirely to the latter isotope. Samples of air containing radon can readily be obtained from an apparatus known as a 'thorium cow'. This consists of a flexible plastic bottle containing a thorium compound which produces radon as it decays. By squeezing the bottle the air containing radon can be transferred to any desired vessel.

Apparatus and Materials

Part (a)
- Scaler with thin-walled G–M tube, holder, and lead
- 'Thorium cow' connected to a beaker, with a cork to take G–M tube as shown in Figure 3.3
- Stop-clock (or automatic timing control)

Part (b)
- Electrometer d.c. amplifier, with ionization chamber connected to suitable meter and 'thorium cow' as shown in Figure 3.4

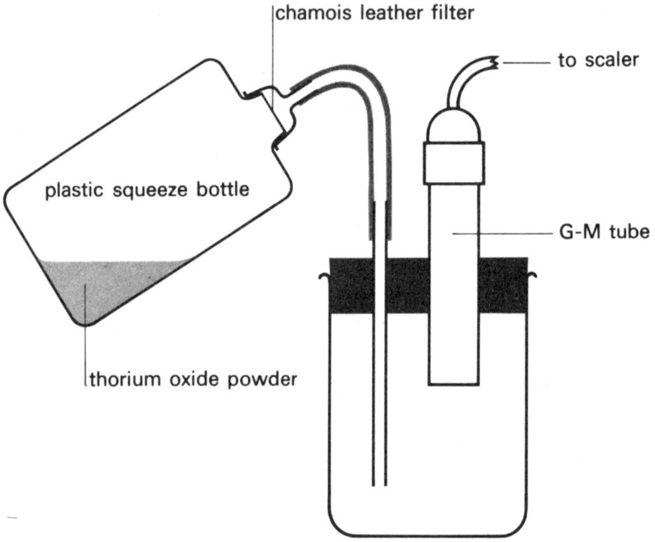

Figure 3.3 Investigating the decay of radon-220, procedure (a)

Procedure

(a) Squeeze the 'thorium cow' bottle to introduce radon gas into the beaker. As the sample decays, count it over twenty second intervals with ten seconds rest between each, as in Experiment 3.1. Plot a graph to show the rate of decay and from it obtain the half-life of radon-220.

Although a thin-walled G–M tube is used, a large proportion of the α-particles emitted are absorbed and thus uncounted. The count rates are therefore low and the accuracy of the determination is correspondingly reduced. A second method, described below, overcomes the problem of α-particle absorption by injecting the radioactive gas directly into the counting chamber.

(b) Squeeze the 'thorium cow' to introduce radon gas into the ionization chamber. Use this as a pilot to find the meter range that gives nearest to full-

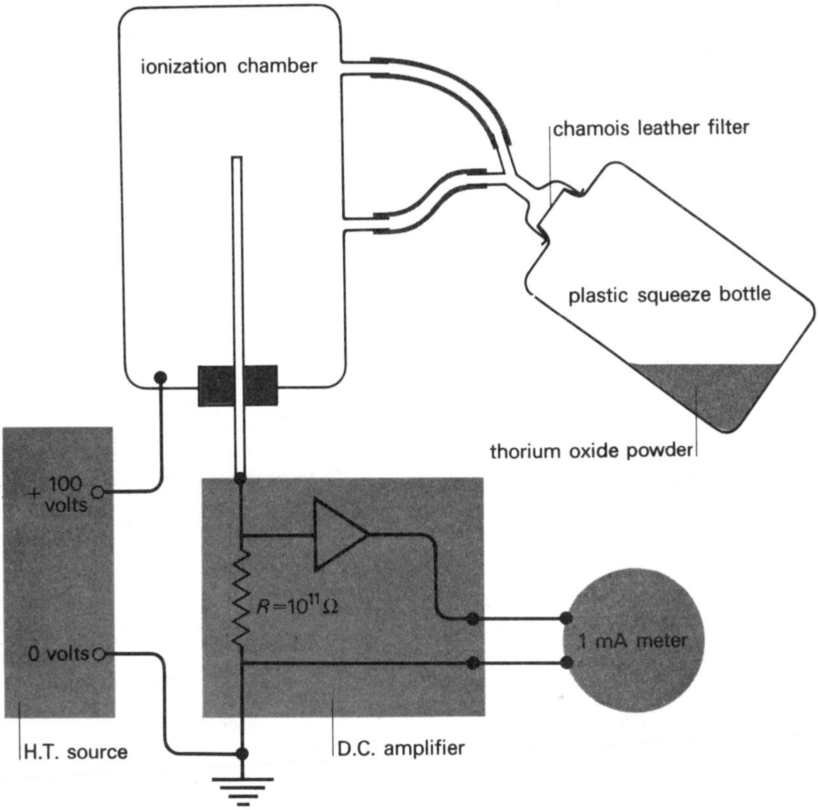

Figure 3.4 Investigating the decay of radon-220, procedure (b)

scale deflection without going beyond the scale. Now introduce a further sample of radon, note the meter reading and simultaneously start the stop-clock. Record the meter readings at half-minute intervals until the reading drops to about one-tenth of the full-scale deflection.

The α-particles cause ionization of the air molecules and a current will flow through the ionization chamber. This current causes a potential difference across the resistor and this, after amplification, is registered on the meter. The meter reading is thus proportional to the number of radon atoms present.

Plot a graph of meter reading against time and from it obtain the half-life of radon-220.

3.3 Separating Lead-212 and Finding its Half-life

In this experiment two methods of separating lead-212 are described. The first uses chemical separation based on the differing solubilities of hydroxides, while the second uses an electrochemical method.

Apparatus and Materials

Part (a)
- Scalar with liquid counter, holder, and lead
- Centrifuge
- 0.10 M thorium(IV) nitrate solution
- 0.2 M lead nitrate and barium nitrate solutions
- 0.880 ammonia solution
- Concentrated nitric acid
- Dilute (2 M) nitric acid

Part (b)
- Scalar with G–M tube, holder, and lead
- Dilute (2 M) ammonia solution
- Metal can, fitted as in Figure 3.5
- Source of d.c. (100–150 volts)
- Planchette
- Centrifuge with tubes
- Infra-red heating lamp
- 0.10 M thorium(IV) nitrate solution

Procedure

(a) Place 10 cm^3 of the thorium(IV) nitrate solution in a centrifuge tube and add a few drops each of the lead nitrate and barium nitrate solutions. Add 0.880 ammonia solution drop by drop until no further precipitation of hydroxides occurs. The precipitate is a mixture of thorium(IV) and lead(II) hydroxides. The barium ions are to prevent the precipitation of radium hydroxide.

Warm the liquid until the precipitate settles, then cool, and centrifuge. Discard the supernatant liquid and redissolve the precipitate in the least possible quantity of dilute nitric acid. Reprecipitate the hydroxides with 0.880 ammonia, centrifuge, and wash with hot water, discarding the rinsings in each case. Dissolve the precipitate, which is now reasonably free from adsorbed ions, in 5 cm^3 of dilute nitric acid. Place the solution in the liquid counter and count it for a five-minute period. Repeat the counting at hourly intervals (or thereabouts) for the next twenty-four hours.

(b) Fill the centrifuge tube to a depth of 2 cm with thorium(IV) nitrate solution and add ammonia solution dropwise until no further precipitation occurs. Centrifuge the mixture and discard the supernatant liquid. Carefully transfer the residue of thorium(IV) hydroxide to the planchette and dry it thoroughly under an infra-red heating lamp where it will be converted to thorium(IV) oxide. A freshly prepared source of thorium such as this is essential so that the first daughter product, radon-228, is readily freed in the gaseous state. Transfer the planchette to the apparatus shown in Figure 3.5, complete the assembly, and leave it connected to the electrical supply for four hours. The can will become filled with gaseous ions of the entire

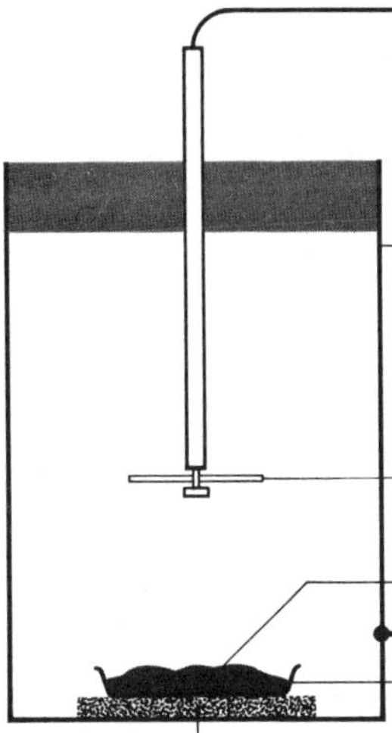

Figure 3.5 The electrical separation of lead-212

range of daughter products of the thorium decay series (see Figure 3.2) mixed with the air. Ions of lead-212, with oxidation number +2, will be deposited almost exclusively on the cathode during this time by the reaction:

$$^{212}Pb^{2+}(g) + 2e^- \longrightarrow {}^{212}Pb(s)$$

Disconnect the electrical supply, remove the cathode, and put it on a planchette under the G–M tube. Count its activity over a five-minute period. Repeat the counting at hourly intervals (or thereabouts) over the next twenty-four hours.

Results

(a) and (b) Plot the count rate against time for each method and from the graphs obtain the half-life of lead-212.

3.4 Separating Bismuth-212 and Finding its Half-life

This is another chemical separation process based on the precipitation of bismuth as sulphide.

Apparatus and Materials

Scaler with liquid counter, holder, and lead
Centrifuge
Hydrogen sulphide generator
0.1 M solution of thorium(IV) nitrate
0.2 M solutions of lead nitrate and bismuth nitrate
Concentrated nitric acid
Dilute sulphuric acid

Procedure

Place about 20 cm^3 of the thorium(IV) nitrate solution in a centrifuge tube, add about 2 cm^3 of lead nitrate solution and a few drops of bismuth nitrate solution. Add dilute sulphuric acid until precipitation of lead sulphate is complete. The bismuth ions in the solution act as a 'holdback' carrier for the bismuth-212. Centrifuge and discard the precipitate of lead sulphate. Pass hydrogen sulphide into the liquid you have separated to precipitate the bismuth as sulphide. Again centrifuge and discard the supernatant liquid. Redissolve the bismuth sulphide in concentrated nitric acid and then add a few drops of water to dilute the solution. Place the solution in the liquid counter and count over one minute. Recount the solution at five-minute intervals for up to two hours.

Results

Record the count rates and the times at which they were taken. Plot count rate against time and determine the half-life as in previous experiments.

3.5 Separating Thallium-208 and Finding its Half-life

This chemical separation involves the removal of thallium-208 with ammonium phosphomolybdate which acts as an ion exchange resin (or molecular sieve).

Apparatus and Materials

Scaler with G–M tube, holder, and lead
Demountable filtration apparatus (as shown in Figure 3.6) with filter pump or line
Filter paper (Whatman's No. 1) to fit above
$\frac{1}{4}$ of a Whatman's accelerator (1″ × $\frac{3}{4}$″)
Planchette
50 per cent solution of thorium(IV) nitrate made up in 2 M nitric acid
Ammonium phosphomolybdate (prepared as described in the Appendix)

Procedure

Finely shred the portion of Whatman's accelerator and disperse it with 0.15 g of ammonium phosphomolybdate in 5 cm^3 of water. Filter this through

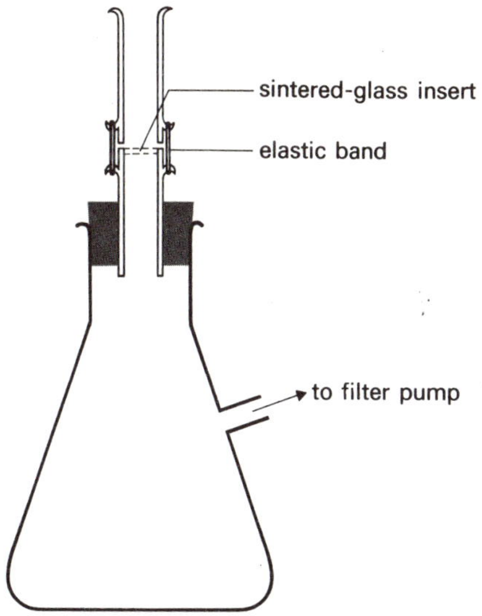

Figure 3.6 The separation of thallium-208

the demountable filtering apparatus containing the Whatman's No. 1 filter paper, using a moderate reduction of pressure. *Without allowing the solid to dry out* run 10 cm^3 of the thorium(IV) nitrate solution through the filtration apparatus, adjusting the pressure so that this takes about two minutes. Pass 10 cm^3 of dilute nitric acid through the assembly at a similar rate and suck dry.

Transfer the ammonium phosphomolybdate, which will have selectively retained the thallium-208, to a planchette. Count for thirty seconds. Recount at thirty second intervals for about fifteen minutes.

Results

Record the count rate and the time.

Plot count rate against time and from the graph obtain the half-life of thallium-208.

4: Some Tracer Techniques

Just as naturalists, wishing to study the migration patterns of birds, find it useful to 'tag' some specimens, scientists obtain help in understanding the mechanism of chemical reactions by 'labelling' some of the atoms taking part in reactions by using radioactive isotopes. This enables them to follow the course of these atoms during the reaction. In the experiments which follow you will meet several examples of the use of this technique.

4.1 Is there an Equilibrium between Solid Lead Chloride and its Saturated Aqueous Solution?

One of the most important applications of radioactive tracers is to the investigation of equilibria. Before the advent of this method it had always been assumed that in a saturated solution there was a dynamic equilibrium between the salt in solution and the solid remaining undissolved. By using 'tagged' molecules or ions it is now possible to find out whether this is true or not. (See also Experiment 1.1 in Book 2 (Electrochemistry).)

Apparatus and Materials

Scaler with liquid counter, holder, and lead
Stop-clock
Concentrated hydrochloric acid
Apparatus for filtration under pressure
Planchette
Infra-red heating lamp
Centrifuge and tubes
5 cm^3 pipette
1 M solutions of lead nitrate and thorium(IV) nitrate
2 M hydrochloric acid

Procedure

Prepare a saturated solution of lead chloride by mixing 5 cm^3 each of 1 M lead nitrate solution and 2 M hydrochloric acid. Cool the mixture, filter, and retain the filtrate.

Prepare some solid lead chloride containing some lead-212 by mixing together 5 cm^3 each of 1 M lead nitrate solution, 1 M thorium(IV) nitrate solution, and 2 M hydrochloric acid. Filter and wash the residue.

Place the saturated lead chloride solution in the liquid counter and find the background count over a suitable time interval. Place this solution in a centrifuge tube and add one-quarter of the solid lead chloride containing some lead-212. Cork the tube and shake the contents for about five minutes.

Centrifuge the mixture and place the supernatant liquid in the liquid counter. Count its activity. Has the activity changed? What does this tell you about the possibility of a dynamic equilibrium between the solid and its saturated solution?

An additional experiment

It is possible, using this technique, to determine the solubility product of lead chloride (see Book 2).

Weigh a planchette, place the remaining radioactive lead chloride on it, and dry it under the infra-red lamp. Reweigh the planchette and dissolve *all* the lead chloride it carries in 10 cm^3 of concentrated hydrochloric acid. Calculate the molarity of this solution. Place the solution in the liquid counter and count it. Use the count rate of this solution of known molarity to calculate the molarity of lead(II) ions in the saturated solution previously counted, and hence obtain the solubility product of lead chloride.

4.2 Investigation of the Reaction between Solutions of Thorium(IV) Nitrate and Sodium Oxalate

In this experiment you will be able to follow a chemical reaction involving a radioactive substance. You should see that you understand exactly what you are doing before you start and how the method adopted gives you the information you want.

Apparatus and Materials

2 burettes
Centrifuge and tubes
Scaler, liquid counter, and associated apparatus
Stop-clock
Portion of Whatman's accelerator
0.1 M solutions of thorium(IV) nitrate and sodium oxalate

Procedure

Prepare the mixtures of the two solutions specified in Table 4.1. As the precipitate that forms is partially colloidal, each mixture should be boiled for a few minutes with fragments of Whatman's accelerator to complete the precipitation. Separate each mixture using a centrifuge. Place each 'filtrate' in turn in the liquid counter and determine the count rate in counts per minute. The counting time for each sample should be varied so that about 500 counts are recorded on the scaler. Record your results in Table 4.1.

Results

Table 4.1

Volume of 0.1 *M thorium nitrate in* cm^3	0	1	2	3	4	5	6	7	8	9	10
Volume of 0.1 *M sodium oxalate in* cm^3	10	9	8	7	6	5	4	3	2	1	0
Count rate in min^{-1}											

Plot a graph of count rate against volume composition of the mixture and from the graph deduce:

(a) the ionic equation for the reaction between thorium nitrate and sodium oxalate and

(b) the solubility produce of the solid precipitated.

4.3 Investigation of the Distribution of Iodine between Tetrachloromethane and Water

Equilibria of this kind have been dealt with by conventional methods of analysis in Book 1. In this case, the analysis can be very conveniently carried out by using 'tagged' iodine and finding the distribution coefficient from the activities of the two layers.

Apparatus and Materials

Scaler with liquid counter, holder, and lead

Tablet preparation of iodine-131 (this is potassium iodide with a proportion of $^{131}I^-$ ions)

Separating funnel (about 200 cm^3 capacity)

Pair of dissection forceps

Pair of scissors

150 cm^3 beakers (2)

100 cm^3 measuring cylinder

Tetrachloromethane

Dilute sulphuric acid

20 volume hydrogen peroxide solution

Procedure

(*Note: the vapour of tetrachloromethane is toxic.*) Hold the packet containing the iodine tablet with the forceps, snip off the corner, and drop the tablet into 25 cm^3 of dilute sulphuric acid in a beaker. When it has completely dissolved, add 25 cm^3 of hydrogen peroxide. This oxidizes the iodide ions to iodine molecules. Transfer the solution to the separating funnel, add

25 cm^3 of tetrachloromethane, and shake the mixture together. Much of the iodine will pass into the tetrachloromethane layer. Run this off and reserve it for the time being. Add a further 25 cm^3 portion of tetrachloromethane to the aqueous layer, shake for a few minutes, and then run off the organic layer and add it to that obtained in the first extraction. The aqueous layer now contains very little iodine and can be disposed of.

Rinse the separating funnel with water, and put the iodine solution in tetrachloromethane into the funnel. Add 50 cm^3 of water and shake these together. Allow the layers to separate and run the two layers into separate beakers, discarding the small amount between the two layers (in the radioactive waste bottle). Count each layer separately. Replace both layers in the separating funnel, add 25 cm^3 more of water, and repeat the shaking, separation, and counting. Repeat the addition of 25 cm^3 portions of water and subsequent treatment until there is a total of 175 cm^3 of liquid in the separating funnel.

Results

Complete the table below with the corresponding count rates for the two layers.

Activity of tetrachloromethane layer in count min^{-1}					
Activity of aqueous layer in count min^{-1}					

Plot a graph of the results you have obtained and use this to determine the relationship between the concentrations in the two layers.

Questions

1. Why was it necessary to carry out the preliminary separation as detailed in the first paragraph of the procedure?
2. Consider the advantages and disadvantages of the radiochemical method as opposed to the conventional method of carrying out this determination.

4.4 Finding the Equilibrium Constant of a Redox Reaction

It is possible to find equilibrium constants of certain reactions by radiochemical methods provided suitable isotopes can be obtained for 'tagging' purposes. In the present experiment the reaction to be studied is

$$Fe^{2+}(aq) + Ag^{+}(aq) \rightleftharpoons Fe^{3+}(aq) + Ag(s)$$

There is a silver isotope, ^{110}Ag, which has a suitable half-life (253 d) and can be used for this purpose. The activity of the unprecipitated silver can be determined, from which its concentration can be obtained, as well as those of Fe^{2+}(aq) and Fe^{3+}(aq).

Apparatus and Materials

Scaler with liquid counter, holder, and lead
Small glass-stoppered bottle (about 25 cm³ capacity)
Tablet preparation of silver-110
250 cm³ beaker
200 cm³ measuring flask
5 cm³ pipette
Centrifuge and tubes
10 cm³ pipette
0.20 M solutions of iron(II) sulphate and barium nitrate
Analar silver nitrate

Procedure

(*Note: radioactive solutions must not be pipetted by mouth.*) Prepare a 0.10 M solution of silver nitrate by the following method. Open the envelope containing the ^{110}Ag preparation by the method described in the last experiment. Place the tablet in a previously weighed beaker and reweigh. Add 200 cm³ of water and sufficient Analar silver nitrate to make the total mass of silver nitrate (including that of the tablet) 0.340 g. The water used must, of course, be distilled or deionized. Prepare a 0.10 M solution of iron(II) nitrate by pipetting 10 cm³ each of 0.20 M iron(II) sulphate and 0.20 M barium nitrate into a centrifuge tube, separating and discarding the residue.

Place 10 cm³ each of the 0.10 M silver nitrate and iron(II) nitrate solutions in a small glass-stoppered bottle, and leave it overnight in a dark cupboard for the equilibrium to be established. Taking care not to disturb the precipitate of silver, remove a 10 cm³ sample of the supernatant liquid and count it. Find also the count rate of the 0.10 M silver nitrate solution.

Results and Calculation

Record the count rates and from them calculate:
(a) the molarity of silver ions in the equilibrium mixture,
(b) the molarity of iron(II) ions at equilibrium,
(c) the molarity of iron(III) ions at equilibrium, and
(d) the equilibrium constant, K_c.

Questions

1. If you carried out the experiments in Book 2 you will have determined this equilibrium constant by an electrochemical method and by a method involving volumetric analysis. How do your results compare?
2. Which method do you consider to be (a) the most convenient and (b) the most accurate?

4.5 Investigation of the Uptake of Radioactive Materials by Plants

Radiochemical methods are very useful in tracing the path taken by substances in a plant or animal body, and they are widely used in biological and medical research. In this experiment we shall find out where phosphorus and/or sulphur go when taken up by a leaf.

Apparatus and Materials

Tablet preparation of phosphorus-32 and/or tablet preparation of sulphur-35. If both elements are to be used the apparatus listed below will be required in duplicate
50 cm^3 beaker
Split cork, bored to accept stems of twigs
Retort stand
Ratemeter with G–M tube, holder, and lead
Cellophane bag and piece of white card large enough to be used to mount a detached leaf
Wrapped X-ray film large enough to cover the leaf
Dark room with accessories to develop X-ray film (see Experiment 1.1)
Freshly cut leafy twigs (laurel is suitable)

Procedure

Set up the experiment as shown in Figure 4.1. The beaker should contain 10 cm^3 of the radioactive solution made by dissolving the tablet preparation

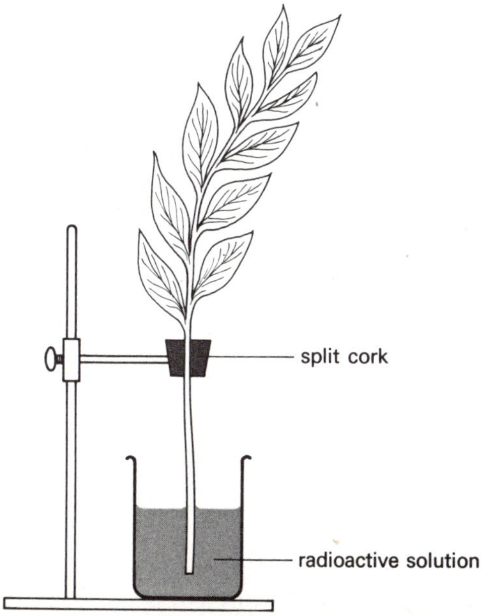

Figure 4.1 Uptake of radioactive ions by plants

in about 30 cm^3 of water. Use the ratemeter to monitor the progress of the radioactive material up the twig and to the leaves by placing the G–M tube at different heights up the twig. When a leaf shows good and fairly regular activity, detach it from the twig and mount it on the white card. Then put the mounted leaf inside the cellophane bag and leave it in contact with the wrapped X-ray film undisturbed in the dark for about a week. If a clear photograph is to be obtained the bag must not, of course, be moved relative to the film during this time. Develop the X-ray film (see Experiment 1.1) and examine it carefully. What parts of the leaves have acquired the greatest activity?

4.6 Investigation of the Separation of Sugars Using Paper Chromatography

This experiment, strictly speaking, is not a radiochemical one, but it is introduced here to give you some practice in a method of separation which we shall use in the following experiment. If you have worked through the experiments in Book 3 of this series (*Chemical Analysis, Chromatography, and Ion Exchange*) you will be familiar with the technique of paper chromatography, which is used here. The results of this experiment are required to evaluate those of Experiment 4.7.

Apparatus and Materials

Chromatography tank with paper of suitable size
Melting point tubes
100 cm^3 measuring cylinder
Oven at 100 °C
1 per cent solutions of glucose, fructose, maltose, and sucrose
1 per cent solution of aniline in propanone (acetone)
2 per cent solution of diphenylamine in propanone
85 per cent aqueous solution of orthophosphoric acid
Solvent containing butan-1-ol, ethanoic acid, and water in the volume proportions 4:1:2

Procedure

(Note: use a graphite pencil for all markings on chromatography paper.) Draw a base line about 2 cm from the bottom of a sheet of chromatography paper which will go into the chromatography tank. Leaving margins at either side of at least 2 cm, mark four equidistant origins on this line, labelling each with the name of one of the sugars to be run. By means of the melting point tubes apply one drop of each of the sugar solutions to its correct origin and allow the spots to dry thoroughly. Place the solvent in the chromatography tank and allow the chromatogram to run until it is a few centimetres below the top of the paper. Remove the paper from the tank and carefully mark on it the height to which the solvent rises; this is called the solvent front. Hang the paper until it is thoroughly dry (a hair dryer may be used to accelerate the process if one is available). Meanwhile prepare the locating agent by mixing the aniline, diphenylamine, and orthophosphoric acid solutions in the volume proportions 5:5:1. Dip the paper in the locating agent and place it in the hot oven for between two and three minutes. The colour produced on the paper by reaction between the sugar and the locating agent may fade rapidly and should therefore be ringed with pencil immediately. Put a dot in the centre of each ring and calculate the R_f value for each sugar. The R_f value is defined as

$$R_f = \frac{\text{distance test substance travels}}{\text{distance solvent front travels}}$$

and is constant for a given substance whenever the chromatogram is run under the same conditions. Figure 4.2 shows you how these distances should be measured.

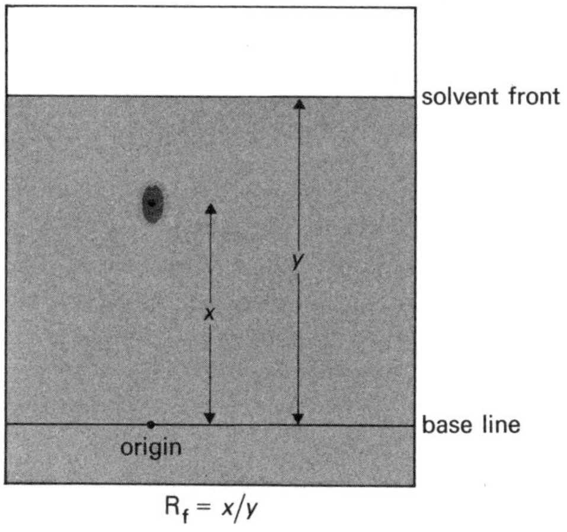

Figure 4.2 The chromatographic separation of sugars

4.7 Investigation of the Photosynthetic Activity of a Plant using Carbon-14 as a Tracer

This experiment is a good example of the use of a radiochemical technique to study a fundamentally important biological problem. It uses ^{14}C to fix the positions of the places reached by sugars in a chromatogram and hence to obtain their R_f values from which they can be identified.

Apparatus and Materials

Leak-proof, clear polythene bag of 30 dm^3 capacity
Healthy geranium plant
Rubber bung fitted with L-bend glass tubes, rubber connections, and screw-clips as in Figure 4.3
Tablet preparation of carbon-14
50 cm^3 beaker
5 cm^3 plastic disposable syringe
Filter pump
Ratemeter with G–M tube, holder, and connectors
Chromatography tank with paper of suitable size
Melting point tube
X-ray film (wrapped) of same size as chromatography paper
Dark room containing developing and fixing materials as required for development of the X-ray film
Concentrated hydrochloric acid
Wash-bottle containing 2 M potassium hydroxide

Procedure

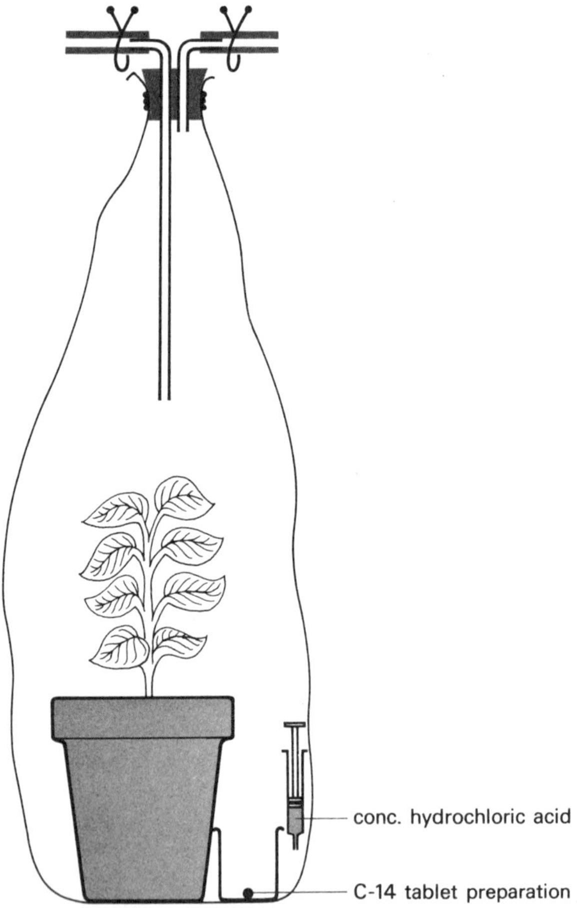

Figure 4.3 Investigating the photosynthetic activity of plants using carbon-14

Set up the apparatus as shown in Figure 4.3. The tablet preparation of carbon-14 is calcium carbonate containing the radioactive isotope so that when acid is added to it carbon dioxide containing $^{14}C^{16}O_2$ is released. Transfer the hydrochloric acid from the syringe into the beaker containing the tablet to liberate the active carbon dioxide into the polythene bag. Place the entire apparatus in a position where it can receive direct sunlight, and leave it there for some days. The carbon-14 will have been taken up by the plant from the carbon dioxide present, and it will be present in any sugars formed during the plant's photosynthetic activity. Connect the filter pump via the wash-bottle to the apparatus shown in Figure 4.4. Open the clips and draw the air from the apparatus through the potassium hydroxide solution for several minutes to absorb the excess carbon dioxide. *The solution in the wash-bottle should be carefully disposed of as it contains carbon-14 of half-life 5745 years*!

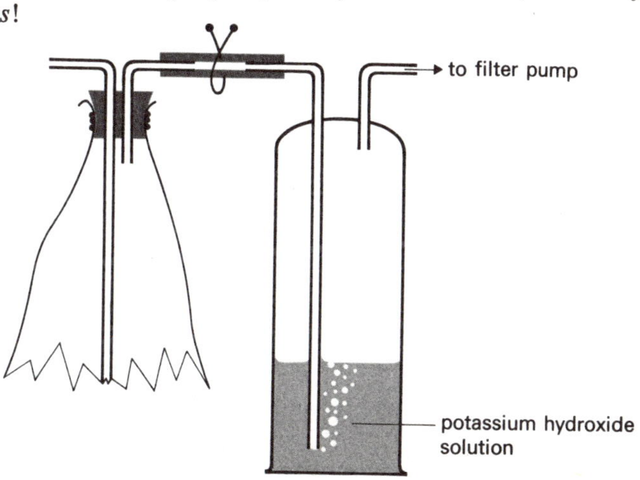

Figure 4.4 Removal of excess radioactive carbon dioxide

Remove the plant from the bag and use the ratemeter to test the uptake of radioactive material by the plant. Remove a leaf from the plant and cut it into very fine pieces. Boil these fragments for two or three minutes with the minimum quantity of water. Prepare a piece of chromatography paper as in Experiment 4.5, drawing the base line as before. To the base line apply a drop of the water in which the pieces of leaf have been boiled and run the chromatogram as in the last experiment. Remove the chromatogram from the tank and allow it to dry. Place it over the wrapped X-ray film, keeping them firmly in contact by means of paper clips, and with a ball-point pen and ruler score over the base line and the solvent front so that an impression of these is made on the X-ray film. Leave the chromatogram in contact with the X-ray film for about two weeks. Develop the film, and calculate the R_f values of any spots that are formed. From these values and those obtained in Experiment 4.5 for known sugars, identify the sugars that have been produced by photosynthesis.

Appendix

The Radioactive Decay Series

Thorium Series (4*n*)

- ^{232}Th → (α, 1.39×10^{10} y) → ^{228}Ra
- ^{228}Ra → (β, 6.7 y) → ^{228}Ac
- ^{228}Ac → (β, 6.13 h) → ^{228}Th
- ^{228}Th → (α, 1.90 y) → ^{224}Ra
- ^{224}Ra → (α, 3.64 d) → ^{220}Rn
- ^{220}Rn → (α, 54.5 s) → ^{216}Po
- ^{216}Po ($\alpha + \beta$, 0.16 s): ~100% → ^{212}Pb; 0.014% → ^{216}At
- ^{212}Pb → ($\beta + \alpha$, 10.6 h) → ^{212}Bi
- ^{216}At → (α, 3×10^{-4} s) → ^{212}Bi
- ^{212}Bi ($\beta + \alpha$, 60 m): 66.3% → ^{212}Po; 33.7% → ^{208}Tl
- ^{212}Po → (α, 3×10^{-7} s) → ^{208}Pb
- ^{208}Tl → (β, 3.1 m) → ^{208}Pb

Neptunium Series (4*n*+1)

- ^{241}Pu → (β, 10 y) → ^{241}Am
- ^{241}Am → (α, 500 y) → ^{237}Np
- ^{237}Np → (α, 2.2×10^{6} y) → ^{233}Pa
- ^{233}Pa → (β, 27.4 d) → ^{233}U
- ^{233}U → (α, 1.62×10^{5} y) → ^{229}Th
- ^{229}Th → (α, 7.0×10^{3} y) → ^{225}Ra
- ^{225}Ra → (β, 14.8 d) → ^{225}Ac
- ^{225}Ac → (α, 10.0 d) → ^{221}Fr
- ^{221}Fr → (α, 4.8 m) → ^{217}At
- ^{217}At → (α, 0.018 s) → ^{213}Bi
- ^{213}Bi ($\beta + \alpha$, 47 m): 96% → ^{213}Po; 4% → ^{209}Tl
- ^{213}Po → (α, 4.2×10^{-6} s) → ^{209}Pb
- ^{209}Tl → (β, 2.2 m) → ^{209}Pb
- ^{209}Pb → (β, 3.3 h) → ^{209}Bi

Uranium Series (4*n*+2)

- ^{238}U → (α, 4.5×19 y) → ^{234}Th
- ^{234}Th → (β, 24.1 d) → ^{234}Pa
- ^{234}Pa → (β, 1.14 m) → ^{234}U
- ^{234}U → (α, 2.35×10^{5} y) → ^{230}Th
- ^{230}Th → (α, 8.0×10^{4} y) → ^{226}Ra
- ^{226}Ra → (α, 1.62×10^{3} y) → ^{222}Rn
- ^{222}Rn → (α, 3.82 d) → ^{218}Po
- ^{218}Po ($\beta + \alpha$, 3 m): 99.96% → ^{214}Pb; 0.04% → ^{218}At
- ^{214}Pb → (β, 26.8 m) → ^{214}Bi
- ^{218}At → (α, 2 s) → ^{214}Bi
- ^{214}Bi ($\beta + \alpha$, 19.7 m): 99.96% → ^{214}Po; 0.04% → ^{210}Tl
- ^{214}Po → (α, 1.5×10^{-4} s) → ^{210}Pb
- ^{210}Tl → (β, 1.32 m) → ^{210}Pb
- ^{210}Pb → (β, 22 y) → ^{210}Bi
- ^{210}Bi ($\beta + \alpha$, 5 d): ~100% → ^{210}Po; ~10^{-5}% → ^{206}Tl
- ^{210}Po → (α, 140 d) → ^{206}Pb
- ^{206}Tl → (β, 4.23 m) → ^{206}Pb

Actinium Series (4*n*+3)

- ^{235}U → (α, 7.07×10^{8} y) → ^{231}Th
- ^{231}Th → (β, 24.6 h) → ^{231}Pa
- ^{231}Pa → (α, 3.2×10^{4} y) → ^{227}Ac
- ^{227}Ac ($\beta + \alpha$), 21.7 y): 98.8% → ^{227}Th; 1.2% → ^{223}Fr
- ^{227}Th → (α, 18.9 d) → ^{223}Ra
- ^{223}Fr → (β, 21 m) → ^{223}Ra
- ^{223}Ra → (α, 11.2 d) → ^{219}Rn
- ^{219}Rn → (α, 3.92 s) → ^{215}Po
- ^{215}Po ($\beta + \alpha$, 1.8×10^{-3} s): ~100% → ^{211}Pb; ~5×10^{-4}% → ^{215}At
- ^{211}Pb → (β, 3.61 m) → ^{211}Bi
- ^{215}At → (α, 10^{-4} s) → ^{211}Bi
- ^{211}Bi ($\beta + \alpha$, 2.16 m): 99.68% → ^{211}Po; 0.32% → ^{207}Tl
- ^{211}Po → (α, 5×10^{-3} s) → ^{207}Pb
- ^{207}Tl → (β, 4.76 m) → ^{207}Pb

The Preparation of Ammonium-12-Molybdophosphate

Apparatus and Materials

Grade 2 sintered-glass filter funnel, Buchner flask, and filter pump
25 cm^3 measuring cylinder
2 cm^3 graduated pipette
100 cm^3 beaker
Desiccator
Watch-glass
Storage bottle
10 per cent aqueous solution of ammonium molybdate
Ammonium nitrate
Concentrated nitric acid
20 per cent w/v aqueous orthophosphoric acid

Procedure

Measure 25 cm^3 of the ammonium molybdate solution into the beaker and raise the temperature to about 40 °C. Add and dissolve 4 g of ammonium nitrate and 1.5 cm^3 of concentrated nitric acid. Add 7 cm^3 of orthophosphoric acid solution and cool the mixture to room temperature. Allow the mixture to stand for some minutes to complete the precipitation of ammonium-12-molybdophosphate and then filter through the sintered-glass funnel. Wash the precipitate with distilled water and draw air through it to dry it as much as possible. Transfer the precipitate to the watch-glass and place in the desiccator for several hours before use.

Logarithms

	0	1	2	3	4	5	6	7	8	9	1	2	3	4	5	6	7	8	9
10	0000	0043	0086	0128	0170	0212	0253	0294	0334	0374	4	8	12	17	21	25	29	33	37
11	0414	0453	0492	0531	0569	0607	0645	0682	0719	0755	4	8	11	15	19	23	26	30	34
12	0792	0828	0864	0899	0934	0969	1004	1038	1072	1106	3	7	10	14	17	21	24	28	31
13	1139	1173	1206	1239	1271	1303	1335	1367	1399	1430	3	6	10	13	16	19	23	26	29
14	1461	1492	1523	1553	1584	1614	1644	1673	1703	1732	3	6	9	12	15	18	21	24	27
15	1761	1790	1818	1847	1875	1903	1931	1959	1987	2014	3	6	8	11	14	17	20	22	25
16	2041	2068	2095	2122	2148	2175	2201	2227	2253	2279	3	5	8	11	13	16	18	21	24
17	2304	2330	2355	2380	2405	2430	2455	2480	2504	2529	2	5	7	10	12	15	17	20	22
18	2553	2577	2601	2625	2648	2672	2695	2718	2742	2765	2	5	7	9	12	14	16	19	21
19	2788	2810	2833	2856	2878	2900	2923	2945	2967	2989	2	4	7	9	11	13	16	18	20
20	3010	3032	3054	3075	3096	3118	3139	3160	3181	3201	2	4	6	8	11	13	15	17	19
21	3222	3243	3263	3284	3304	3324	3345	3365	3385	3404	2	4	6	8	10	12	14	16	18
22	3424	3444	3464	3483	3502	3522	3541	3560	3579	3598	2	4	6	8	10	12	14	15	17
23	3617	3636	3655	3674	3692	3711	3729	3747	3766	3784	2	4	6	7	9	11	13	15	17
24	3802	3820	3838	3856	3874	3892	3909	3927	3945	3962	2	4	5	7	9	11	12	14	16
25	3979	3997	4014	4031	4048	4065	4082	4099	4116	4133	2	3	5	7	9	10	12	14	15
26	4150	4166	4183	4200	4216	4232	4249	4265	4281	4298	2	3	5	7	8	10	11	13	15
27	4314	4330	4346	4362	4378	4393	4409	4425	4440	4456	2	3	5	6	8	9	11	13	14
28	4472	4487	4502	4518	4533	4548	4564	4579	4594	4609	2	3	5	6	8	9	11	12	14
29	4624	4639	4654	4669	4683	4698	4713	4728	4742	4757	1	3	4	6	7	9	10	12	13
30	4771	4786	4800	4814	4829	4843	4857	4871	4886	4900	1	3	4	6	7	9	10	11	13
31	4914	4928	4942	4955	4969	4983	4997	5011	5024	5038	1	3	4	6	7	8	10	11	12
32	5051	5065	5079	5092	5105	5119	5132	5145	5159	5172	1	3	4	5	7	8	9	11	12
33	5185	5198	5211	5224	5237	5250	5263	5276	5289	5302	1	3	4	5	6	8	9	10	12
34	5315	5328	5340	5353	5366	5378	5391	5403	5416	5428	1	3	4	5	6	8	9	10	11
35	5441	5453	5465	5478	5490	5502	5514	5527	5539	5551	1	2	4	5	6	7	9	10	11
36	5563	5575	5587	5599	5611	5623	5635	5647	5658	5670	1	2	4	5	6	7	8	10	11
37	5682	5694	5705	5717	5729	5740	5752	5763	5775	5786	1	2	3	5	6	7	8	9	10
38	5798	5809	5821	5832	5843	5855	5866	5877	5888	5899	1	2	3	5	6	7	8	9	10
39	5911	5922	5933	5944	5955	5966	5977	5988	5999	6010	1	2	3	4	5	7	8	9	10
40	6021	6031	6042	6053	6064	6075	6085	6096	6107	6117	1	2	3	4	5	6	8	9	10
41	6128	6138	6149	6160	6170	6180	6191	6201	6212	6222	1	2	3	4	5	6	7	8	9
42	6232	6243	6253	6263	6274	6284	6294	6304	6314	6325	1	2	3	4	5	6	7	8	9
43	6335	6345	6355	6365	6375	6385	6395	6405	6415	6425	1	2	3	4	5	6	7	8	9
44	6435	6444	6454	6464	6474	6484	6493	6503	6513	6522	1	2	3	4	5	6	7	8	9
45	6532	6542	6551	6561	6571	6580	6590	6599	6609	6618	1	2	3	4	5	6	7	8	9
46	6628	6637	6646	6656	6665	6675	6684	6693	6702	6712	1	2	3	4	5	6	7	7	8
47	6721	6730	6739	6749	6758	6767	6776	6785	6794	6803	1	2	3	4	5	5	6	7	8
48	6812	6821	6830	6839	6848	6857	6866	6875	6884	6893	1	2	3	4	4	5	6	7	8
49	6902	6911	6920	6928	6937	6946	6955	6964	6972	6981	1	2	3	4	4	5	6	7	8
50	6990	6998	7007	7016	7024	7033	7042	7050	7059	7067	1	2	3	3	4	5	6	7	8
51	7076	7084	7093	7101	7110	7118	7126	7135	7143	7152	1	2	3	3	4	5	6	7	8
52	7160	7168	7177	7185	7193	7202	7210	7218	7226	7235	1	2	2	3	4	5	6	7	7
53	7243	7251	7259	7267	7275	7284	7292	7300	7308	7316	1	2	2	3	4	5	6	6	7
54	7324	7332	7340	7348	7356	7364	7372	7380	7388	7396	1	2	2	3	4	5	6	6	7
	0	**1**	**2**	**3**	**4**	**5**	**6**	**7**	**8**	**9**	**1**	**2**	**3**	**4**	**5**	**6**	**7**	**8**	**9**

	0	1	2	3	4	5	6	7	8	9	1	2	3	4	5	6	7	8	9
55	7404	7412	7419	7427	7435	7443	7451	7459	7466	7474	1	2	2	3	4	5	5	6	7
56	7482	7490	7497	7505	7513	7520	7528	7536	7543	7551	1	2	2	3	4	5	5	6	7
57	7559	7566	7574	7582	7589	7597	7604	7612	7619	7627	1	2	2	3	4	5	5	6	7
58	7634	7642	7649	7657	7664	7672	7679	7686	7694	7701	1	1	2	3	4	4	5	6	7
59	7709	7716	7723	7731	7738	7745	7752	7760	7767	7774	1	1	2	3	4	4	5	6	7
60	7782	7789	7796	7803	7810	7818	7825	7832	7839	7846	1	1	2	3	4	4	5	6	6
61	7853	7860	7868	7875	7882	7889	7896	7903	7910	7917	1	1	2	3	4	4	5	6	6
62	7924	7931	7938	7945	7952	7959	7966	7973	7980	7987	1	1	2	3	3	4	5	6	6
63	7993	8000	8007	8014	8021	8028	8035	8041	8048	8055	1	1	2	3	3	4	5	5	6
64	8062	8069	8075	8082	8089	8096	8102	8109	8116	8122	1	1	2	3	3	4	5	5	6
65	8129	8136	8142	8149	8156	8162	8169	8176	8182	8189	1	1	2	3	3	4	5	5	6
66	8195	8202	8209	8215	8222	8228	8235	8241	8248	8254	1	1	2	3	3	4	5	5	6
67	8261	8267	8274	8280	8287	8293	8299	8306	8312	8319	1	1	2	3	3	4	5	5	6
68	8325	8331	8338	8344	8351	8357	8363	8370	8376	8382	1	1	2	3	3	4	4	5	6
69	8388	8395	8401	8407	8414	8420	8426	8432	8439	8445	1	1	2	2	3	4	4	5	6
70	8451	8457	8463	8470	8476	8482	8488	8494	8500	8506	1	1	2	2	3	4	4	5	6
71	8513	8519	8525	8531	8537	8543	8549	8555	8561	8567	1	1	2	2	3	4	4	5	5
72	8573	8579	8585	8591	8597	8603	8609	8615	8621	8627	1	1	2	2	3	4	4	5	5
73	8633	8639	8645	8651	8657	8663	8669	8675	8681	8686	1	1	2	2	3	4	4	5	5
74	8692	8698	8704	8710	8716	8722	8727	8733	8739	8745	1	1	2	2	3	4	4	5	5
75	8751	8756	8762	8768	8774	8779	8785	8791	8797	8802	1	1	2	2	3	3	4	5	5
76	8808	8814	8820	8825	8831	8837	8842	8848	8854	8859	1	1	2	2	3	3	4	5	5
77	8865	8871	8876	8882	8887	8893	8899	8904	8910	8915	1	1	2	2	3	3	4	4	5
78	8921	8927	8932	8938	8943	8949	8954	8960	8965	8971	1	1	2	2	3	3	4	4	5
79	8976	8982	8987	8993	8998	9004	9009	9015	9020	9025	1	1	2	2	3	3	4	4	5
80	9031	9036	9042	9047	9053	9058	9063	9069	9074	9079	1	1	2	2	3	3	4	4	5
81	9085	9090	9096	9101	9106	9112	9117	9122	9128	9133	1	1	2	2	3	3	4	4	5
82	9138	9143	9149	9154	9159	9165	9170	9175	9180	9186	1	1	2	2	3	3	4	4	5
83	9191	9196	9201	9206	9212	9217	9222	9227	9232	9238	1	1	2	2	3	3	4	4	5
84	9243	9248	9253	9258	9263	9269	9274	9279	9284	9289	1	1	2	2	3	3	4	4	5
85	9294	9299	9304	9309	9315	9320	9325	9330	9335	9340	1	1	2	2	3	3	4	4	5
86	9345	9350	9355	9360	9365	9370	9375	9380	9385	9390	1	1	2	2	3	3	4	4	5
87	9395	9400	9405	9410	9415	9420	9425	9430	9435	9440	0	1	1	2	2	3	3	4	4
88	9445	9450	9455	9460	9465	9469	9474	9479	9484	9489	0	1	1	2	2	3	3	4	4
89	9494	9499	9504	9509	9513	9518	9523	9528	9533	9538	0	1	1	2	2	3	3	4	4
90	9542	9547	9552	9557	9562	9566	9571	9576	9581	9586	0	1	1	2	2	3	3	4	4
91	9590	9595	9600	9605	9609	9614	9619	9624	9628	9633	0	1	1	2	2	3	3	4	4
92	9638	9643	9647	9652	9657	9661	9666	9671	9675	9680	0	1	1	2	2	3	3	4	4
93	9685	9689	9694	9699	9703	9708	9713	9717	9722	9727	0	1	1	2	2	3	3	4	4
94	9731	9736	9741	9745	9750	9754	9759	9763	9768	9773	0	1	1	2	2	3	3	4	4
95	9777	9782	9786	9791	9795	9800	9805	9809	9814	9818	0	1	1	2	2	3	3	4	4
96	9823	9827	9832	9836	9841	9845	9850	9854	9859	9863	0	1	1	2	2	3	3	4	4
97	9868	9872	9877	9881	9886	9890	9894	9899	9903	9908	0	1	1	2	2	3	3	4	4
98	9912	9917	9921	9926	9930	9934	9939	9943	9948	9952	0	1	1	2	2	3	3	4	4
99	9956	9961	9965	9969	9974	9978	9983	9987	9991	9996	0	1	1	2	2	3	3	3	4
	0	**1**	**2**	**3**	**4**	**5**	**6**	**7**	**8**	**9**	**1**	**2**	**3**	**4**	**5**	**6**	**7**	**8**	**9**

e^{-x} from $x = 0.00$ to $x = 5.50$ (in increments of 0.01)

x	e^{-x}	x	e^{-x}	x	e^{-x}	x	e^{-x}	x	e^{-x}
0.00	1.0000	**0.55**	0.5770	**1.10**	0.3329	**1.65**	0.1921	**2.20**	0.1108
0.01	0.9901	0.56	5712	1.11	3296	1.66	1901	2.21	1097
0.02	9802	0.57	5655	1.12	3263	1.67	1882	2.22	1087
0.03	9704	0.58	5599	1.13	3230	1.68	1864	2.23	1075
0.04	9608	0.59	5543	1.14	3198	1.69	1845	2.24	1065
0.05	0.9512	**0.60**	0.5488	**1.15**	0.3166	**1.70**	0.1827	**0.25**	0.1054
0.06	9418	0.61	5434	1.16	3135	1.71	1809	2.26	1044
0.07	9324	0.62	5379	1.17	3104	1.72	1791	2.27	1033
0.08	9231	0.63	5326	1.18	3073	1.73	1773	2.28	1023
0.09	9139	0.64	5273	1.19	3042	1.74	1755	2.29	1013
0.10	0.9048	**0.65**	0.5220	**1.20**	0.3012	**1.75**	0.1738	**2.30**	0.1003
0.11	8958	0.66	5169	1.21	2982	1.76	1720	2.31	0993
0.12	8869	0.67	5117	1.22	2952	1.77	1703	2.32	0983
0.13	8781	0.68	5066	1.23	2923	1.78	1686	2.33	0973
0.14	8694	0.69	5016	1.24	2894	1.79	1670	2.34	0963
0.15	0.8607	**0.70**	0.4966	**1.25**	0.2865	**1.80**	0.1653	**2.35**	0.0954
0.16	8521	0.71	4916	1.26	2837	1.81	1637	2.36	0944
0.17	8437	0.72	4868	1.27	2808	1.82	1620	2.37	0935
0.18	8353	0.73	4819	1.28	2780	1.83	1604	2.38	0926
0.19	8270	0.74	4771	1.29	2753	1.84	1588	2.39	0916
0.20	0.8187	**0.75**	0.4724	**1.30**	0.2725	**1.85**	0.1572	**2.40**	0.0907
0.21	8106	0.76	4677	1.31	2698	1.86	1557	2.41	0898
0.22	8025	0.77	4630	1.32	2671	1.87	1541	2.42	0889
0.23	7945	0.78	4584	1.33	2645	1.88	1526	2.43	0880
0.24	7866	0.79	4538	1.34	2618	1.89	1511	2.44	0872
0.25	0.7788	**0.80**	0.4493	**1.35**	0.2592	**1.90**	0.1496	**2.45**	0.0863
0.26	7711	0.81	4449	1.36	2567	1.91	1481	2.46	0854
0.27	7634	0.82	4404	1.37	2541	1.92	1466	2.47	0846
0.28	7558	0.83	4360	1.38	2516	1.93	1451	2.48	0837
0.29	7483	0.84	4317	1.39	2491	1.94	1437	2.49	0829
0.30	0.7408	**0.85**	0.4274	**1.40**	0.2466	**1.95**	0.1423	**2.50**	0.0821
0.31	7334	0.86	4232	1.41	2441	1.96	1409	2.51	0813
0.32	7261	0.87	4190	1.42	2417	1.97	1395	2.52	0805
0.33	7189	0.88	4148	1.43	2393	1.98	1381	2.53	0797
0.34	7118	0.89	4107	1.44	2369	1.99	1367	2.54	0789
0.35	0.7047	**0.90**	0.4066	**1.45**	0.2346	**2.00**	0.1353	**2.55**	0.0780
0.36	6977	0.91	4025	1.46	2322	2.01	1340	2.56	0773
0.37	6907	0.92	3985	1.47	2299	2.02	1327	2.57	0765
0.38	6839	0.93	3946	1.48	2276	2.03	1313	2.58	0758
0.39	6771	0.94	3906	1.49	2254	2.04	1301	2.59	0750
0.40	0.6703	**0.95**	0.3867	**1.50**	0.2231	**2.05**	0.1287	**2.60**	0.0743
0.41	6637	0.96	3829	1.51	2209	2.06	1275	2.61	0735
0.42	6570	0.97	3791	1.52	2187	2.07	1262	2.62	0728
0.43	6505	0.98	3753	1.53	2165	2.08	1249	2.63	0721
0.44	6440	0.99	3716	1.54	2144	2.09	1237	2.64	0714
0.45	0.6376	**1.00**	0.3679	**1.55**	0.2122	**2.10**	0.1225	**2.65**	0.0707
0.46	6313	1.01	3642	1.56	2101	2.11	1212	2.66	0699
0.47	6250	1.02	3606	1.57	2080	2.12	1200	2.67	0693
0.48	6188	1.03	3570	1.58	2060	2.13	1188	2.68	0686
0.49	6126	1.04	3535	1.59	2039	2.14	1177	2.69	0679
0.50	0.6065	**1.05**	0.3499	**1.60**	0.2019	**2.15**	0.1165	**2.70**	0.0672
0.51	6005	1.06	3465	1.61	1999	2.16	1153	2.71	0665
0.52	5945	1.07	3430	1.62	1979	2.17	1142	2.72	0659
0.53	5886	1.08	3396	1.63	1959	2.18	1130	2.73	0652
0.54	5827	1.09	3362	1.64	1940	2.19	1119	2.74	0646